THE AUTOMOBILE

By Michael Burgan

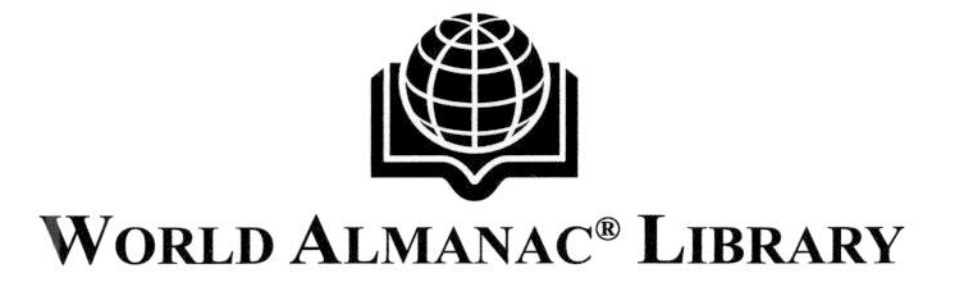

WORLD ALMANAC® LIBRARY

Please visit our web site at: www.worldalmanaclibrary.com
For a free color catalog describing World Almanac® Library's list of high-quality books and multimedia programs, call 1-800-848-2928 (USA) or 1-800-387-3178 (Canada). World Almanac® Library's fax: (414) 332-3567.

Library of Congress Cataloging-in-Publication Data

Burgan, Michael.
The automobile / by Michael Burgan.
p. cm. — (Great inventions)
Includes bibliographical references and index.
ISBN 0-8368-5800-X (lib. bdg.)
1. Automobiles—Juvenile literature. I. Title.
TL146.5.B87 2005
629.222—dc22 2004056930

First published in 2005 by
World Almanac® Library
330 West Olive Street, Suite 100
Milwaukee, WI 53132 USA

A Creative Media Applications, Inc. Production
Design and Production: Alan Barnett, Inc.
Editors: Matt Levine, Susan Madoff
Copy Editor: Laurie Lieb
Proofreader: Tania Bissell
Indexer: Nara Wood
World Almanac® Library editorial direction: Mark J. Sachner
World Almanac® Library editor: Gini Holland
World Almanac® Library art direction: Tammy West
World Almanac® Library production: Jessica Morris

Photo credits: © AP/Wide World Photos: pages 6, 19, 21, 22, 23, 24, 25, 27, 29, 31, 33, 34, 35, 36, 37, 38, 39, 41, 42, 43; © Bettmann/CORBIS: pages 5, 9, 26; © CORBIS: page 18; © Getty Images/Hulton Archive: page 12; © Hulton-Deutsch Collection/CORBIS: page 16; © North Wind Picture Archives: pages 4, 7, 11, 14; diagram by Rolin Graphics: page 13.

Printed in Canada

1 2 3 4 5 6 7 8 9 09 08 07 06 05

TABLE OF CONTENTS

Words that appear in the glossary are printed in **boldface** type the first time they appear in the text.

CHAPTER 1 THE FIRST "HORSELESS CARRIAGES"

Imagine a world without automobiles. To travel on land, people would have to use their own power or ride animals. No autos would also mean a loss of freedom and choice. Cars let people live miles from where they work, shop, or go to school. Without cars, people would have to live closer to the important places and people in their lives.

Cars provide jobs, as well. Many important industries have developed because of the auto. Around the world, millions of people make auto parts, build cars, and keep them running. Others find, produce, and sell the gasoline that powers automobiles. People who travel long distances by car need places to sleep and

▼ *Before the automobile was invented, people traveled from place to place on foot, on horseback, or in horse-drawn carriages such as those shown here.*

▲ *This picture, taken in 1906, shows the Panhard automobile, one of the earliest cars manufactured for everyday needs.*

eat. Hotels, motels, and restaurants hire millions of workers to care for car travelers' needs.

Cars can also provide entertainment. Each year, millions of people attend auto races, and many more watch them on television. Lovers of **antique** autos go to shows just to admire the beauty and rarity of cars from the past.

Cars, trucks, and other road vehicles have met people's desire to get from one location to another as easily as possible. They have improved life in many ways.

The Era of Animal Power

For thousands of years, humans relied on their feet to get from one place to another when traveling on land. The first great improvement in transportation came more than fifty-five hundred years ago. Humans in Asia and Africa began to tame wild animals and train them to carry people's goods. The first of these pack animals was an early ancestor of the donkey. Some people then started riding animals. The most useful of these animals was the horse. Fast

Fast Fact

Horses were first tamed about six thousand years ago in Central Asia. They were used as a source of food before humans began riding them.

The da Vinci Car

Leonardo da Vinci (1452–1519) was a great Italian painter. He was also a scientist and engineer who filled books with drawings of many inventions. He designed an airplane, a helicopter, and a tank centuries before they were actually built. Da Vinci also drew what may have been the world's first **self-propelled** car. It looked like a wooden tricycle, and its power came from small, compressed springs. When the springs were released, they created energy that turned the car's wheels. Da Vinci did not leave behind a working model of his car. In 2004, however, an Italian museum used his drawings to create a small version of the car. A scientist who helped build the car compared it to the **robotic** vehicles used to explore Mars. Both are small and self-propelled. Both also can be set to move in a certain direction before they start to roll.

and strong, horses could cover long distances and carry soldiers into battle.

More than five thousand years ago, humans began to develop the wheel. This invention, combined with pack animals, opened up a new world of transportation. The first wheels were made of solid pieces of wood tied together. Slowly, early engineers developed spoked wheels. The spokes made the wheels lighter while keeping them strong enough to support heavy loads. People attached wheels to carts and used animals, such as oxen, to pull the carts. The animals could haul heavier items when the goods were in carts than if the goods rested on their backs. Most early carts had just two wheels. Later, four-wheeled carts, or wagons, became popular. Carts designed to carry people were often called carriages. In battle, carts pulled by horses were called chariots.

About A.D. 500, Chinese engineers began to build sail cars. Just like sailboats, these wagons relied on wind power. One Chinese sail car could carry thirty people at speeds of about 30 miles (48 kilometers) per hour. The sail car, however, was not perfect because it could be used only on windy days. Inventors began to dream of vehicles that could carry their source of power with them.

The Steam Age

The ancient Greeks and Romans knew that steam created by boiling water could be a source of power. The Greek scientist Hero (c. A.D. 100) used rushing steam to turn a small metal ball. Hero's invention, however, was just a toy. Scientists and engineers did not create practical steam engines until the eighteenth century.

The English inventor Thomas Newcomen (1663–1729) perfected a steam engine in 1712. Later in the century, James Watt (1736–1819) developed a more powerful steam engine that was soon used across Europe and North America. At first, steam engines were used mostly in mines to pump out water that often flooded the bottom of the mines. By 1800, steam engines were also appearing in factories. These engines replaced waterwheels as the source of power for the factories' large machines.

Some inventors also realized that steam power could replace horses for propelling vehicles. In 1769, a Frenchman named Nicolas-Joseph Cugnot (1728–1804) built what is considered the first working self-propelled vehicle. His invention was

Fast Fact

In 1784, James Watt installed pipes carrying steam in his office. This was the first use of steam heat.

◀ *This diagram shows how the Greek scientist Hero used steam to spin a toy globe. Hot water in the cauldron (A–B) creates steam, which then flows up through pipes (L and G) and fills the globe. Pipes K and H allow steam to escape, causing the globe to spin on its axis.*

called a road locomotive. It had three large wheels and was designed to pull cannons for the French army.

In a steam engine, steam presses against a metal rod called a **piston**, making it move up and down. This movement creates power. Cugnot used another metal rod called a crankshaft to carry the power from the piston to the wheels of his vehicle. The turning crankshaft moved the wheels. At the time, steam engines were large and heavy, so Cugnot's vehicle moved very slowly.

After Cugnot, inventors began to focus on using steam to power ships and locomotives that pulled wagons on rails. Before the end of the eighteenth century, the first steamships appeared in both Europe and the United States. The first steam locomotive rolled down a set of tracks in England in 1804.

About 1805, U.S. inventor Oliver Evans (1755–1819) tested a steam vehicle that traveled on

Evans and the Steam Engine

Oliver Evans produced a more powerful steam engine than the one commonly used at the end of the eighteenth century. In 1805, he outlined his ideas for future steam-powered vehicles. Envisioning trucks, trains, and steamboats, here is part of what he said:

> I conceive that carriages may be constructed, to be propelled by the power of steam engines which I have invented, to transport merchandise and produce from Philadelphia to Columbia, and from thence to Philadelphia, much cheaper than can be done by the use of cattle.... This carriage I allow to carry 100 barrels of flour, and to travel 3 miles [4.8 km] per hour on level roads, and 1 mile [1.6 km] per hour up and down hills; say about 40 miles [64.3 km] per 24 hours; making a trip from Columbia to Philadelphia in 2 days. It requires 5 horse wagons...to transport 100 barrels the same distance in 3 days.... I have no doubt but that my engines will propel boats against the current of the Mississippi, and wagons on turnpike roads with great profit.

both land and water. Evans called it the *Orukter Amphibolos*, Greek for "amphibious digger." His invention was the first self-propelled vehicle to drive on land in the United States. Evans successfully drove the digger down the streets of Philadelphia, Pennsylvania, and into the water.

▲ *The first steam engines were large and heavy and moved very slowly.*

During the 1820s and 1830s, several British engineers built large, steam-powered carriages. As on steam trains, coal was burned to heat water inside a boiler. The boiling water produced steam. Belts or chains transferred the power created in the steam engine to the rear wheels. A driver sat in front and steered the vehicle by turning a metal rod connected to the front wheels. The steam carriages were noisy and smoky, and they scared horses and angered people who shared the road with them. By the 1860s, the British government passed laws that limited how fast steam-powered vehicles could travel. Local officials also won the power to limit the hours that steam buses could use their roads. These laws prevented the further development of

Watt Wasn't Horsing Around

While working on his steam engine, James Watt did not forget the importance of the horse to humans. He called the unit of measure for his engine's power the **horsepower** (hp). To power a mill used to grind grain, horses typically walked in a circle about 75.4 feet (22.9 meters) around. Watt figured that the average horse walked that distance 144 times in one hour, reaching a speed of about 181 feet (55 m) per minute. Watt then guessed that the horse pulled with a force of 180 pounds (82 kilograms). He multiplied these two numbers and came up with a figure close to 33,000. One horsepower equals 33,000 foot-pounds (44,750 joules) per minute. In other words, lifting 33,000 pounds (15,000 kg) of weight 1 foot (0.3 m) in one minute requires 1 horsepower of energy. Today, a lawn mower engine might have five horsepower, while a car has more than one hundred.

steam buses. Still, in both Europe and North America, inventors built new steam vehicles. The first self-propelled fire engines used steam power, and small steam cars appeared on U.S. roads.

Going Electric

During the 1820s and 1830s, scientists in Europe and the United States studied electricity. They learned that electricity had a special relation with magnets. Electricity could make a metal coil act as a magnet, creating what was called an **electromagnet**. The scientists also learned that they could create electricity by moving a magnet in and out of a metal coil. With this knowledge, a Vermont blacksmith named Thomas Davenport (1802–1851) built a simple electric motor in 1833. During the 1840s, several inventors built cars powered with electric motors that ran on batteries. None of these were successful.

Inventors also tried to build electric vehicles that traveled on railroads. One early electric locomotive

Fast Fact

The large British steam carriages were actually the first buses—they carried passengers who paid to travel between cities.

built about 1850 had a motor that produced 16 horsepower. The batteries required for this motor, however, were large and heavy. Engineers got around this problem by designing electric trains and **trolleys** that did not use batteries. The vehicles took their power from electric lines above the tracks or electricity that flowed through a third rail. To build electric road cars, however inventors first had to develop smaller batteries. Cars could not run on electric rails or be easily connected to power lines, as trains were.

The world's first public electric railway opened in Germany in 1881. The first electric cars, however, appeared about 1890. Several U.S. inventors created them. One of the most successful was a large tricycle built by William Morrison (c. 1850–1927) of Iowa. Compared to steam cars, electric cars were clean and quiet. They were also fast. Electric cars were a popular type of "horseless carriage." But the need to recharge batteries still limited how long a person could drive an electric car. The development of the modern automobile rested on engines powered by another fuel: gasoline.

▼ *The Babcock Electric Victoria Phaeton was one of the earliest electric cars. Its use, however, was limited to the short amount of time the battery stayed charged.*

CHAPTER 2

GOING WITH GAS

During the late seventeenth century, a Dutch scientist named Christiaan Huygens (1629–1695) experimented with gunpowder as fuel for an engine. In theory, the gunpowder would burn inside a round metal tube called a cylinder. The burning fuel would create a gas, which would expand and push against a piston inside the cylinder. Huygens never built a working engine using his design, and gunpowder was too dangerous to use as a fuel. But the idea of combusting, or burning, a fuel inside a cylinder led to the **internal combustion** engine.

▼ *Étienne Lenoir's combustion engine, which used a mixture of air and coal gas as fuel, created an explosion that took place outside of the engine, unlike later, more advanced internal combustion engines.*

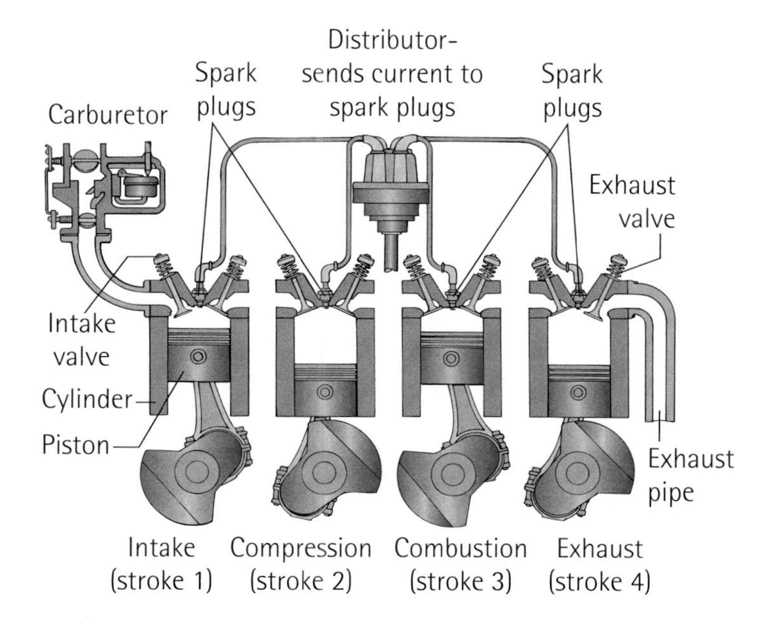

◄ *In a typical internal combustion engine of an automobile, gasoline is burned inside the engine, releasing an enormous amount of energy, which is then converted into motion in a 4-stroke (4-step) cycle. On the first stroke, each piston begins at the top of its cylinder. As the piston moves down, it sucks in air and fuel that has been mixed in the carburetor through an intake valve. On the second stroke, the piston moves upward to compress the fuel. On the third stroke, an electric spark (the result of an electric current sent through the distributor to the spark plugs) combusts the fuel. The resulting hot gases push the piston back down with great force. On the fourth and last stroke, the piston moves back to the top of its cylinder and pushes the gases out through the exhaust valve.*

Several inventors tried to use combustion in engines they designed. During the mid-nineteenth century, this coal gas was often used for cooking and light inside homes. In 1859, a Belgian mechanic named Étienne Lenoir (1822–1900) built a working engine that used a mixture of air and coal gas as fuel. An electric spark combusted the fuel, creating the gas that moved the piston.

Lenoir's engine was not truly an internal combustion engine. The explosion that created the hot gas took place outside of the cylinder that held the piston. French inventor Alphonse Beau de Rochas (1815–1893) suggested how to build the first true internal combustion engine. The fuel, he said, should be **ignited** inside the cylinder. Beau de Rochas also improved on Lenoir's engine by compressing the fuel before igniting it. With this process, his engine used less fuel.

Beau de Rochas never built his engine, but he outlined the four separate steps, or strokes, the piston would go through to create power (*see diagram*).

A New Fuel

During the 1860s, a German engineer named Nikolaus Otto (1832–1891) began working on internal combustion engines. By 1876, he had perfected a four-stroke engine. He and a business partner sold the engines to factories. Like Lenoir's engine, the Otto engine ran on coal gas. A number of inventors, however, believed that a liquid fuel would make more sense for powering a car, because such a fuel could be carried on the vehicle. An internal combustion engine that was powered by coal gas had to be connected to a pipe that carried the gas.

The liquid fuel most inventors tried was called benzine. Today, it is better known as gasoline. This fuel comes from petroleum, a thick, black substance found underground. Petroleum formed millions of years ago. As plants and animals died, they were covered with rock and sand. Over time, heat and pressure in the earth turned the dead plants and animals into petroleum. Humans have known about petroleum for thousands of years. In some places, it is close to the surface and bubbles up through holes in the ground. Most petroleum is buried deep beneath the ground. In 1859, a U.S. company began drilling into the earth to find petroleum. By this time, scientists had shown that petroleum could be separated into different useful substances. This process is called **refining**. One key petroleum product was kerosene, which could be used as a fuel in lamps. Another product of the refining process was gasoline.

▼ *In the mid-1800s, the United States began drilling beneath the ground for oil in an effort to find efficient fuel for internal combustion engines.*

The Unknown Auto Inventor

Many books on the history of the automobile ignore Siegfried Marcus (1831–1898) of Austria. He was possibly the greatest inventor whom almost no one knows. Marcus worked with electricity and chemicals, as well as with internal combustion engines and automobiles. By 1865, he had perfected a simple gasoline-powered car, and within ten years, he had built three more cars. One could travel at about 4 miles (6.4 km) per hour. Marcus, however, grew tired of experimenting with cars and moved on to other projects.

In the late 1940s, workers at a museum in Vienna, Austria, found one of Marcus's cars from the 1870s. Even after sitting unused for decades, it still ran. This car is the world's oldest existing gasoline-powered vehicle.

For a time, scientists and engineers thought that gasoline was useless—and dangerous, since it exploded easily. Several inventors, however, thought that the ease of combusting gasoline would make it a perfect fuel for internal combustion engines.

Daimler and Benz

Two of these successful inventors were Gottlieb Daimler (1834–1900) and Wilhelm Maybach (1847–1929). They worked for Otto's engine company but left to pursue their own engine design. Daimler thought that gasoline was the ideal fuel for engines that could power road vehicles. He and Maybach used a device called a carburetor to **vaporize** liquid gasoline. As a vapor, the gasoline was easily mixed with air and then ignited. Daimler and Maybach's gasoline engine was lighter and much more powerful than Otto's original engine. In April 1885, Daimler received a **patent** for his new engine. The patent prevented other engineers from copying his design.

Fast Fact

People first used petroleum as a medicine, as fuel for lamps, and to seal cracks in wooden ships.

Fast Fact

Neither Daimler nor Maybach knew how to ride a bike. They added one small wheel on either side of their motorcycle to help them keep their balance.

A few months later, Daimler and Maybach used their engine to power a wooden bicycle, creating the world's first motorcycle. Bikes were still a new invention at this time. Most still had one huge wheel in front, and the rider sat above it. Bikes with two tires of the same size and a seat for the rider between them were just starting to appear. Daimler's teenage son Paul took most of the test rides on the new vehicle.

Next, Daimler took a carriage normally pulled by a horse and added an engine to it. A series of belts connected the engine and the rear wheels. At first, the local police would not let him drive his horseless carriage on the road. When he did go for a drive, Daimler went slowly so he would not attract much attention. His son Paul later wrote, "The carriage ran well and reached a speed of 18 km/h [11 mph]."

During the early 1880s, another German engineer was working on his own gasoline-powered engine. Carl Benz (1844–1929) had long thought about building a road car. He wrote that his vehicle would run "under its own power, like a locomotive, but not

► *Gottlieb Daimler and Wilhelm Maybach attached a gasoline-powered combustion engine to a carriage. The result was a horseless carriage that reached a speed of 11 miles (18 km) per hour.*

on tracks, but like a wagon simply on any street." In 1883, Benz had several partners in a company that built gasoline engines. The partners said he could try to build cars once their company was making money. Two years later, Benz produced a three-wheeled car. A metal handle called a tiller let the driver steer the single front wheel. Most historians consider this the first practical gasoline-powered car. Benz had created a vehicle that ran well and that he could sell to others.

Improving the Gas Car

With their work, Daimler and Benz had proven that gasoline-powered vehicles made sense. (The two men later teamed up and made cars together. Their company exists today and makes the Mercedes Benz, as well as other cars.) Some inventors in other countries made their own internal combustion engines and tried to make cars. Other early carmakers bought Daimler engines and put them in their own cars. In France, Émile Levassor (c. 1844–1897) and René Panhard (1841–1908) used Daimler engines when they started building cars around 1890. Unlike Daimler and Benz, they placed the engine in the front of the car. This weight at the front plus the passengers' weight in the back gave the car better balance than cars with rear

Mrs. Benz Does Her Part

Benz's wife, Bertha, played a part in her husband's early career as an automaker. She is credited with taking the first long-distance ride in a gasoline-powered car. She wanted to prove to Carl that his invention was a success. One morning in 1888, she and her two sons sneaked out of the house and took a 50-mile (80-km) drive. The car lacked enough power to carry them up hills, so Bertha and her oldest son had to push the vehicle. Still, by nightfall, they had reached their destination. Carl Benz then improved his car so it could go up hills under its own power.

▲ *Carl Benz developed the three-wheeled vehicle that most historians consider the first practical gasoline-powered car.*

engines. The Panhard-Levassor cars also had a gearbox, clutch, and transmission behind the engine. These parts worked together to bring power from the engine to the rear wheels. The car's 2-hp engine could reach speeds of 12.4 miles (20 km) per hour. The company warned, however, that such "great speeds require considerable attention on the part of the driver, and are not always advisable."

For the next fifty years, most automakers copied the Panhard-Levassor design. Over time, however, carmakers stopped using the solid rubber, wooden, and metal tires and wheels used on most early bicycles and cars. Instead, they turned to a new invention, the **pneumatic** tire. In 1888, a Scottish doctor named John Dunlop (1840–1921) received a patent for a tire with a rubber tube that was filled with air. The pneumatic tire gave a smoother, quieter ride than solid tires. Adding Dunlop's tire to the car moved automakers one step closer to the modern car of today.

Thanks to the work of Panhard, Levassor, and other French engineers, France became the first major center for making cars. The French gave the world a new word for the vehicles: *automobile*, which means "self-movable." The success of the French cars also made gasoline engines more popular than steam engines and electric motors. Internal combustion engines were safer, quieter, and easier to run than steam engines. The gasoline engines provided more power than electric engines, since getting powerful yet small batteries for cars was still a problem. Steam and electric cars did not disappear right away, but the gasoline-powered car was soon the favorite around the world.

CHAPTER 3

THE GENIUS OF HENRY FORD

In 1888, the design for the Daimler gasoline engine reached the United States, where American inventors and engineers were also building their own gasoline engines. The first successful U.S. makers of gasoline-powered autos were Charles Duryea (1861–1938) and J. Frank Duryea (1869–1967). These brothers built their first car in 1893 and opened the first U.S. auto manufacturing company three years later. About that time, the great inventor Thomas Edison (1847–1931) made a prediction: "It is only a question of a short time when the carriages and trucks in every large city will be run with motors."

By 1900, many more auto companies had appeared. Americans bought just over four thousand cars that year. Automobiles were still largely for the rich. No one did more than Henry Ford (1863–1947) to make the automobile a part of daily life.

▲ Henry Ford revolutionized the American way of life by manufacturing affordable vehicles with interchangeable parts on an assembly line.

Building the "Universal Car"

In 1896, Ford drove his first car, which was powered by an engine he had built at his home during his spare time. After several years of building race cars,

Fast Fact

As a boy, Ford had watched a steam-powered vehicle chug past him on a country road. From then on, he was fascinated with the idea of building self-propelled road vehicles.

he opened the Ford Motor Company in 1903. His goal was to make a car that was light, simple to drive and repair, and cheap to buy. Ford said he would make a "universal car" that anyone could afford. "No man making a good salary will be unable to own one," he said.

In 1908, Ford produced his dream car, the Model T. For $850, customers could buy a car that held five passengers. It was constructed from a special steel that made the car both light and strong. The car sat several feet off the ground, meaning it could travel easily over bumpy, unpaved roads. In his ads, Ford promised that his car had "at least equal value" with any other car available, yet it "sells for several hundred dollars less than the lowest of the rest."

Within five years, Ford had sold more than 250,000 Model Ts. His constant goal was to build the car faster and cheaper, so he could sell it at a lower price. In 1913, Ford developed a new system that helped him reach that goal. He called it mass production, and it soon changed how companies around the world made their products.

In mass production, Ford combined some existing ideas with his own thoughts on manufacturing. First, he built his cars with interchangeable parts. This

Early Fans of Autos

Some of the first Americans to buy cars were doctors, who often rode long distances to see patients who lived far from town. Some doctors realized that it would be cheaper and faster to have cars than to use horses and wagons for these trips. Following their example, other people began to buy cars. The doctors, however, sometimes had problems with their new vehicles. One doctor in 1901 warned that the latest gasoline-powered cars were heavy. He wrote, "If one gets down into a ditch it is practically impossible to do anything without the assistance of quite a gang of men."

meant that all the pistons, for example, were identical. For hundreds of years, all the parts on a carriage had been made by hand. The parts for one might be slightly different from those for another. If a part broke, mechanics could not guarantee that the replacement would be exactly the same. Interchangeable parts made it easier to both make and fix a car. By Ford's time, guns, sewing machines, and clocks were some of the goods that featured interchangeable parts.

Ford and other manufacturers were able to make interchangeable parts because of the improvements in machine tools. These are large machines that make metal parts for other machines. U.S. companies realized during the early nineteenth century that machine tools could cut certain parts more quickly and easily than humans could. By Ford's time, machine tools could make tiny parts and perfectly round cylinders. The tools cut the parts much more precisely than people could.

For mass production, Ford borrowed another old idea. During the 1780s, steam-engine maker Oliver Evans had built an assembly line on which belts carried wheat through a grain mill. Workers no longer had to carry the grain through the mill. Assembly lines were later used to move dead animals through meatpacking plants. Ford used assembly lines to move car parts through his plants. The lines saved the workers time as they built cars.

▲ *An advertisement for Henry Ford's Model T illustrates the car's good looks and refers to its easy serviceability.*

Racing to Better Cars

Early in his automaking career, Henry Ford focused on building race cars. He and other early automakers drew attention to the horseless carriage by racing. Some races took place over long distances on regular roads, while others were held on closed tracks. Ford and others used what they learned from building race cars to build better passenger cars. Today, most automakers still improve passenger cars by building race cars. For example, General Motors (GM) has studied the kinds of accidents that race car drivers suffer. This knowledge helps the company improve the safety of all cars.

Ford's last major idea was to create a division of labor. In the building of early autos, one team of workers assembled an entire car. Instead, Ford had each of his workers do just one job over and over. For example, a worker might tighten one bolt on an engine. The assembly line then carried the engine to another worker, who added another part. Meanwhile, the assembly line delivered another engine to the previous worker, who tightened the same bolt again.

Ford first tested mass production on a part of the Model T called the flywheel magneto. This piece created the spark that combusted fuel in the engine. The magneto had interchangeable parts, and the parts moved along an assembly line. Under the old system, one worker needed about twenty minutes to build a magneto. Using division of labor, a team of workers could build one in about five minutes.

Mass Production at Work

Ford next applied mass production to building the entire Model T. He saw amazing results. Using the same number of workers, his company could build twice as many cars in one day. This increase in production let Ford cut the cost of his cars, so more people could afford to buy them.

In 1914, Ford decided to raise the salary of most of his workers. He paid them five dollars a day. Before, most Ford workers earned less than three dollars a day. Workers in other industries also earned about that much or even less. Ford could afford to pay the high wages, because mass production was boosting his profits. He believed that the high salaries would encourage his employees to work harder and be loyal to the company.

Ford's success with mass production led other companies to copy his methods. Manufacturers saw that they could hire people with few skills and teach them how to do simple tasks over and over. The business owners also saw that mass production let them lower their prices and therefore sell more goods. Mass production made the United States the center of the auto industry. One historian said that the Model T "revolutionized American life."

Still, not everyone liked the effects of mass production. Workers became bored doing the same small task over and over again. Some also thought Ford wanted them to work too fast. Some Ford workers did not like the control the company had over them. Henry Ford believed that since he paid his workers well, he could tell them what to do inside and outside his plant. By the 1930s, Ford's workers wanted to form labor **unions** to improve their working conditions. Ford and other automakers resisted the unions. Finally, new laws made it easier for workers to form unions. If the companies treated them badly, the workers could refuse to work. Today, most Americans who make cars still belong to unions.

▼ *This picture shows workers on an assembly line in Henry Ford's factory in the early 1900s.*

CHAPTER 4 DECADES OF CHANGE

By the end of the 1910s, Henry Ford was one of the most famous people in the world. He sold the Model T on almost every continent and eventually opened factories in South America, Europe, and Australia. As other carmakers copied his methods, other affordable cars soon competed with the Model T. The automobile was changing the world, and it was going through its own changes.

Cars had begun changing even before Ford introduced mass production. Carmakers wanted to make their products faster and more powerful, so they added more cylinders to their engines. Benz's first car had just one cylinder. By 1900, some cars had four cylinders, and the first six-cylinder engine appeared in 1902. The number of cylinders would later grow to as many as sixteen, but four and six were more common. Engine power, however, was not the only concern. To attract buyers, autos had to become safer and easier to drive. Auto engineers made a series of advancements to meet those goals.

▼ *Henry Ford's Model T became the standard by which affordable cars were judged.*

The Road to Safer Cars

Windshields were one of the first improvements to appear on most cars. Putting a piece of glass in front of the driver kept out bugs, flying rocks, and bad weather. The windshield also freed the driver and passenger from wearing goggles and clothes that covered every inch of their bare skin. Side and rear windows did not appear until years later. Cars were still mostly open, though some had cloth roofs that could be put up in bad weather and rolled down when the sun shone.

▲ *As cars became part of the American way of life, carmakers continued to improve their design. This photograph features a 1905 Cadillac Osceola with sealed headlights.*

Early cars had gas or oil lamps on the fronts to provide light. The lights helped drivers see the road at night and let other people see them. Some of the first lamps created their own gas. Water dripped onto the compound calcium carbide, which created the gas acetylene. The driver then lit the gas before starting to travel. Other early lamps burned kerosene. Sealed lights that used electricity for power began appearing around 1910. Bulb makers first had to design lightbulbs that would not break as a car bounced down bumpy roads.

Better brakes were needed as cars became faster. The first cars often had just a single brake on one of the rear wheels. The driver pulled on a lever attached to the brake. A pad on the brake touched the wheel and slowed it down. The harder the driver pulled, the harder the brake pad pressed against the wheel. Later, a foot pedal replaced the hand lever. Braking improved as carmakers put brakes on both rear wheels. The best brakes used round metal parts,

Fast Fact

The price of the Model T fell to $490 in 1913 and to under $300 by the 1920s.

Kettering's Invention

Henry Leland (1843–1932) ran the Cadillac car company from 1903 to 1917. He had a friend who supposedly died from an accident suffered while he was trying to crank a car. Leland told his engineers that "the Cadillac car will kill no more men if we can help it." Cadillac hired Charles Kettering (1876–1958) to fix the problem. Kettering had just formed his own company to explore how to use electricity in cars. He had already found a way to improve the engine ignition using electricity. His self-starter system was one of the greatest auto improvements ever. Kettering also designed the first all-electric cars. In 1912, some Cadillac cars used electricity for lights, the starter, and ignition. Each of these cars also had an electric generator that created power for its batteries.

called drums. The cars' wheels sat on the drums. The brake pads rubbed against the drums to slow down the car.

Around the 1920s, the first **hydraulic** brakes appeared. A hydraulic braking system used a fluid that flowed through a closed system of tubes. Pressing the brake increased the pressure of the fluid. The fluid then pressed against the brake pads, which pressed against the drums. Hydraulic brakes are still used today.

One of the greatest problems with the first cars was starting them. A driver had to turn a large crank attached to the engine. People often had trouble turning the crank, and sometimes the engine backfired, creating a force that could knock down the person turning the crank. One of the greatest engineering advances came in 1911, when the automaker Cadillac introduced the first electric starter. Power from the battery was used to turn a small electric motor that started the engine. Within a few years, almost all cars had the "self-starter." This invention opened a new world of driving for women, as few of them had been strong enough to start cars using the old crank system.

The Impact of Early Autos

The development of automobiles influenced many areas of daily life. In the United States, road building was greatly affected. Before cars appeared, most American roads were just dirt paths for horse-drawn wagons and carriages. The first U.S. cars were designed so that their wheels would fit in the ruts left in the roads by these wagons. In 1904, the U.S. government reported that fewer than 10 percent of the nation's 2.35 million miles (3.8 million km) of roads were "improved." This meant that gravel or some other hard substance covered the dirt, so the roads could be used in any kind of weather. Most of the improved roads were in cities.

▲ *The Lincoln Highway, running from New York City to San Francisco, California, was built in 1918 with private donations and some money from the federal government.*

As more people bought cars, they demanded better roads. At the time, most roads were built by towns. A road might end where it reached the border with the next town. Slowly, states and the U.S. government began spending money to build and improve roads. One of the first major new roads was planned in 1912. Auto-parts maker Carl Fisher (1874–1939) called for a road across the entire United States. He told other business leaders, "Let's build it before we're too old to enjoy it!" Fisher raised donations for the project, and the U.S. government also gave money. The new road was created by linking and improving existing roads and adding new roads where they were needed. It ran from New York City to San Francisco, California, and was named the Lincoln Highway, in honor of President Abraham Lincoln.

Signals for Safety

Traffic lights help keep cars, trucks, and **pedestrians** safe as they use busy city streets. The first traffic light was introduced in London, England, in 1868 to keep horse-drawn carriages moving. It was a gas-powered lantern with red and green signals. A police officer operated the light. In 1920, two new traffic signals appeared in the United States. One of them was patented by Garrett Morgan (1875–1963), an African American inventor and business owner from Cleveland, Ohio. His signal could stop traffic in all directions so pedestrians could cross the street. Morgan's invention was used across the United States and Canada until the development of the modern, three-colored light still used today.

The increase of cars and roads led to safety issues. The first reported traffic accident in the United States took place in 1896. A man driving a Duryea struck a bicyclist. Three years later, a man in New York was struck by a car as he stepped off a **trolley**. He was the first person known to have been killed by an auto. Government leaders soon realized they had to address the safety issues created by cars. In 1901, Connecticut became the first state to set speed limits. A car could go 15 miles (24 km) per hour in the country and 12 miles (19 km) per hour in the city. The typical horse-drawn carriage had an average speed of about 10 miles (16 km) per hour.

Fast Fact

The first cars to have brakes on each rear wheel appeared about 1910. Most cars did not have front brakes until the 1920s.

More Jobs

The car's practical effect was clear. People could travel long distances any time they chose. The automobile also had an important economic effect. Building cars—and new roads—created jobs. So did running the businesses that developed because people could now travel wherever they wanted.

Thanks to the automobile, the petroleum industry grew. U.S. companies searched for new supplies of this resource, which was also called crude oil. After 1900, most of the new oil wells appeared in the West. Spindletop, Texas, became the center of a new petroleum industry after drillers found a large supply of crude oil there. California also had good supplies of oil. Gasoline was now the most important product created from petroleum because electric lights were replacing kerosene lamps. Petroleum also provided a type of oil needed for auto parts. Without this motor oil, the moving parts would break. As petroleum companies produced more gas and oil, drivers paid less for them. Falling gas prices convinced more people to buy cars—which meant more jobs building them. The demand for cars also helped the rubber industry. Rubber was needed to provide tires for new cars and the replacements for ones that blew out or wore out. During the 1920s, 80 percent of the rubber used in the United States went into cars.

Petroleum companies also created jobs as they built the first gas stations. At first, car owners bought their gas at local general stores, which also sold kerosene and other fuels. Gas stations began to appear about 1912. A car did not drive up to a pump to get gas. Instead, a gas station worker rolled a portable pump

▼ At the first gas stations, workers brought the gas pumps to the cars. Ironically, during these early years, gasoline was delivered to the "filling stations" by horse-drawn carts, as seen in this 1902 photograph.

over to the car. By 1929, almost all car owners bought their gas at "filling stations." The stations' workers also performed repairs, cleaned windshields, and checked to make sure the engines had enough oil.

The automobile also created the modern tourist industry in the United States. People could easily reach remote mountains or lakes and travel from one spot to another on the same trip. In the past, only the wealthy could afford to take vacations like these, traveling by train or private carriage. Now travelers could stay cheaply at new campgrounds that opened for motorists. They could eat at restaurants built along roads far from major cities and towns.

People could make money in isolated places by serving the tourists who passed by. The new restaurants also introduced travelers to new foods. During the 1920s, large, national fast-food companies did not exist. The roadside restaurants served whatever the local people ate. Southerners visiting New England might taste clam chowder for the first time. Midwesterners going to Maryland might have their first fresh shellfish.

Along with cars, gasoline-powered trucks began to appear. Companies that transported goods saw that trucks could be cheaper and faster than trains for covering short distances. Trucking companies had to hire drivers, people to load and unload the trucks, and mechanics to fix them.

Of course, as new jobs were created, old ones began to disappear. People who ran horse stables went out of business, and other jobs connected to the care and feeding of horses began to disappear. Trolley lines that ran into rural towns closed. Trains still ran, but fewer people used them for short trips. The automobile was bringing many changes. Over time, some people realized that not all of these changes were good.

Fast Fact

The automobile reduced the number of towns and cities with trolleys, but they did not completely disappear. Trolleys, also called streetcars, can be found today in such cities as San Francisco, New Orleans, and Tampa. The San Francisco system features streetcars built during the 1930s and 1940s.

CHAPTER 5

NEW CHALLENGES

Between 1914 and 1945, the United States and many other countries fought two world wars. In World War I (1914–1918), cars and other self-propelled, gasoline-powered vehicles played important roles. Trucks brought soldiers to the battlefields, and ambulances carried away the wounded. Manufacturers replaced wheels with a track system and added armor and guns to create the first tanks.

During World War II (1939–1945), U.S. carmakers stopped making autos and once again built tanks and other military vehicles. They also built planes and the engines that powered them. Outside Detroit, Michigan, the Ford Motor Company opened what

▼ *An interior view of the Packard Motor Car Company. in Detroit is shown during the 1920s. In the coming years, U. S. automakers rolled out new cars by the hundreds of thousands to meet the growing needs of a changing society.*

The World Comes to the United States

Before World War II, U.S. auto companies made most of the cars sold in the United States. They also owned car companies in other countries. The war gave U.S. automakers another boost. Germany and Japan, then enemies of the United States, were destroyed by the war. Factories that made their cars were bombed into rubble. After the war, however, the United States helped these and other countries rebuild. Volkswagen was one of several German companies that began selling cars in the United States. Daimler Benz and BMW became famous for their well-made—and expensive—cars. Starting in the 1960s, such Japanese companies as Honda, Toyota, and Nissan sent small cars to America. Later, they sold larger cars, trucks, and SUVs. The rise of the Japanese car companies was particularly amazing, since Japan did not start building cars until just before the war. Today, Toyota and Honda make some of the most popular cars sold in the United States.

was then the largest factory in the world. One newspaper said that the factory was "a promise of American greatness." About 70,000 workers made planes inside the plant. By 1944, Ford was producing more than 400 bomber planes every month.

The war made it hard for **civilians** to keep their vehicles running. To make sure the military had enough fuel, the U.S. government limited how much gas people could buy. Rubber for tires and different metals were also **rationed** in this way.

When World War II ended, Americans were eager to buy new cars again. The U.S. population in 1946 was just over 141 million, and it would begin to increase rapidly during the next two decades. The country had slightly more than twenty-five million cars, and half of them were more than ten years old. From 1946 to 1950, Americans went car crazy, buying more than twenty-one million new vehicles. More than ever before, the United States was the auto capital of the world. Once again, the car changed how people lived.

Fast Fact

To help the war effort, some Americans donated their cars' metal bumpers to be recycled into armaments and replaced them with wood.

Moving to the Suburbs

During the nineteenth century, most people lived close to where they worked. Most walked to their jobs (although those who lived on farms with horses could ride). By the end of the century, the average worker might ride a bicycle, subway, train, or trolley. These vehicles let people live farther from their workplaces. Some even lived in new towns that developed outside of the cities where most jobs were located. These towns were called suburbs. In the suburbs, people could own their own homes with lawns, rather than live in crowded city neighborhoods.

The car led to the rapid growth of suburbs in America. People who wanted to escape the noise and crime of crowded cities settled in these new towns. They then **commuted** by car to their jobs. The boom in auto sales after World War II led to even more expansion of the suburbs.

Once again, cars went through several changes. The automatic transmission had been developed during the 1930s. This device meant that people no longer had to press in the clutch and shift gears as they drove. Automatic transmissions became common after World War II.

Drivers could also find some of the comforts of home in their cars. The first car radios had appeared during the 1930s. Air conditioning, which also first appeared in cars during the 1930s, became more widely available during the 1950s.

During the 1950s, U.S. cars became bigger. People wanted room for children, pets, and their belongings when they went shopping or took family vacations. Some Americans also liked to drive large, expensive cars for another reason. The kind of cars people chose reflected how much money they had—or how much they wanted other people to think they had.

Help from the Government

Several U.S. government policies fueled the growth of suburbs and car ownership. Soldiers returning from World War II received loans to buy homes. If they chose, they could also receive free college educations. The veterans who went to college had better chances of finding high-paying jobs than workers without college degrees. These veterans could afford to buy new cars and live in the suburbs. In the 1950s, the U.S. government also built a new system of wide, straight highways, like the one above in Los Angeles, California. These **interstate** roads made it easy for people to live even farther from their jobs and then commute by car. The interstates also made it easy for tourists to reach every corner of the United States.

Still, some Americans bought small cars if they could not afford larger ones. The German car company Volkswagen sold its Beetle. During the 1960s, Japanese automakers entered this small-car market. A few U.S. companies also built small, cheap cars. One of these was Studebaker. Its Lark was a success when it was first sold in 1959. Another small U.S. car was the two-door Nash Metropolitan. Outside North America, people bought small cars because they were easy to drive and park on narrow city streets. Few countries developed suburbs as quickly as the United States did, so more drivers in other countries still lived in the cities.

▼ *Motorists stop for gas and oil on the New Jersey Turnpike, which was completed in 1951. Turnpikes were the model for today's high-speed interstates, which made travel between large cities and smaller towns easier and faster.*

The bigger cars needed large, powerful engines. Large engines require more fuel than small engines. Most Americans, however, could afford the gas. In 1950, the average price of 1 gallon (3.8 liters) of gasoline was twenty-seven cents, and the price was only four cents higher in 1960.

New Concerns

By the early 1960s, cars were an important part of daily life—in the United States and around the world. Still, some people saw that cars also created dangers.

The number of auto accidents each year was about sixteen million, and close to fifty thousand people died each year in car crashes. In 1965, the U.S. Senate investigated auto safety. As reported in *Car and Driver* magazine, records from the U.S. auto companies showed that "almost one in five autos built in the preceding five years had been **defective.**" The same year, a lawyer named Ralph Nader (1934–) published a book called *Unsafe at Any Speed*. Nader accused carmakers of not correcting known safety problems in some cars. Nader said the U.S. government should force the automakers to build safer cars.

In 1966, Congress responded by passing the National Traffic and Motor Vehicle Safety Act. The law called for better protection of car passengers and better brakes. It also forced car companies to publicly announce when they were recalling cars that had problems that affected safety. During a recall, car owners had to take their cars to service stations, where the problem was fixed.

One safety device already used was the seat belt. The first seat belts were introduced during the 1930s, when some doctors added them to their own cars. The belt went across a passenger's lap. Automakers then began offering seat belts in some of their cars. By the 1950s, some states required all cars to have seat belts. During the late 1950s, the Swedish carmaker Volvo introduced an improved seat belt. It crossed a passenger's upper body, as well as the lap. A similar design is still used today. By the 1960s, U.S. automakers had lap belts in the front seats of all their cars. Over the next few years, the U.S. government passed laws requiring improved seat belts in front and rear seats.

In 1969, the U.S. government asked carmakers to introduce some kind of device that would work on its own to prevent injuries during a crash. One solution

Champion of Safety

Ralph Nader, a U.S. presidential candidate in 1996, 2000, and 2004, was an unknown lawyer when he published *Unsafe at Any Speed.* That book established him as a consumer **advocate**—someone who defends shoppers from unsafe or unfair business practices. In his book, Nader said that GM sold its Chevrolet Corvair even though the company knew the car could be difficult to steer. GM had fixed the problem by the time Nader's book appeared. Still, the book raised the argument that car companies should not be allowed to produce cars that were unsafe. Nader believed that the government had to play a larger role in ensuring safety. "The rule of law," he wrote, "should extend to the safety of any product that carries such high risks to the lives of users and bystanders."

was the air bag. In 1952, U.S. inventor John Hetrick (1918–1999) invented the first air bag. When a crash occurred, the bag would automatically inflate in front of the driver and passengers in the front seat. The bag would prevent them from crashing into the dashboard or windshield. Hetrick never actually developed his air bag or put it into a car. At first, carmakers were not interested in this invention. They thought the bags would be too expensive to design. The carmakers also feared that inflating bags could injure children. Over time, however, the carmakers saw that the bags could save many lives. By the end of the 1980s, most U.S. cars had air bags for the driver.

▲ *This 1962 photograph shows a Los Angeles, California, section of the interstate system that links roads and highways throughout the fifty U.S. states. Unfortunately, a by-product of so many cars on the road is the pollution that hangs in the atmosphere.*

During the early 1950s, scientists had learned that some of the chemicals in engine exhaust added to air pollution. These chemicals included the gases carbon monoxide and nitrogen oxide. Some of the chemicals were produced because a car's fuel was not fully combusted in the engine's cylinders. In 1963, carmakers began adding devices that burned more of the fuel. The cars still emitted some pollution, however. In 1970, the U.S. government forced carmakers to find new ways to reduce pollution. One solution was building engines that ran on unleaded gas. The lead in the old gas had contributed to the pollution problem. Carmakers also added a part to autos called the catalytic converter. The converter made some of the chemicals in car exhaust less dangerous.

The Gas Crisis

At one time, the United States produced enough petroleum to meet its needs. After World War II, however, the country relied more heavily on foreign sources of petroleum. In 1973, events halfway around the world affected driving habits in the United States. Saudi Arabia and other countries in the Middle East refused to ship petroleum to the United States. These countries were protesting U.S. policy in the Middle East.

U.S. refiners needed Middle Eastern oil to create gasoline and other products. Since the refiners now produced less gas, the gas that was available cost more. In the mid-1970s, during what came to be known as the Arab oil embargo, American gas stations and drivers faced high gas prices and shortages for the first time. Sales of large cars that used a lot of gas began to fall. In general, cars at the time did not get good gas mileage. In 1973, all GM cars averaged about 12 miles per gallon (5.1 km/l) of gas. Thanks in part to the oil embargo, automakers realized they had to produce cars that got better gas mileage. The U.S. government also called for average gas mileage to reach 20 miles per gallon (8.5 km/l) by 1980, and even higher averages after that.

To get better mileage, U.S. carmakers began making some cars smaller and lighter. They had to compete with European and Japanese carmakers who had already been selling small cars in the United States.

U.S. automakers improved their engines. The goal was to use less gas to create the same amount of power that the old engines had produced. The government also encouraged people to drive less often. People who worked together were told to carpool, or ride together. Some companies paid for workers to take public transportation. Americans, however, still believed that cars were a necessary part of daily life. Concerns over gas mileage fell when petroleum prices began to decline. Carmakers also began to make larger vehicles again. Therefore, the issue of gas prices and how much people drive reappeared in the decades to come.

▲ *This 1973 photograph shows cars lined up along the Palisades Interstate Parkway, near Fort Lee, New Jersey, awaiting their turn at the pump. Short tempers and impatience often led to altercations between citizens who were frustrated by the gas shortages.*

CHAPTER 6

TODAY AND TOMORROW

Cars entered a new era in the 1970s, when the first computers appeared inside them. Carmakers, taking advantage of the shrinking size of computers, used them to improve their autos' performance.

Onboard computers kept track of the engine's operation. Sensors in different parts of the engine sent information to the computer. The computer could then perform certain functions, such as changing the mixture of gas and air entering the cylinders. In this way, the computer allowed the car to burn less gas and run smoother.

▲ *In the future, computers in cars may warn of dangers in the road and come equipped with sensors to avoid collisions.*

Today, cars feature even more powerful onboard computers, which can control such things as the car's temperature and the information displayed on the dashboard. Auto mechanics also use computers to help them find and fix problems in cars.

The New Family Cars

Other developments from the 1980s and 1990s are found in today's automotive world. One popular passenger vehicle, the minivan, was introduced in

1983 by U.S. automaker Chrysler. Vans had been used for decades in business. A van usually had just two front seats, a large cargo area in the rear, and back doors to load the cargo area. Vans were more like trucks than cars. Chrysler wanted its minivan to drive like a car, yet have more room for passengers and cargo. Chrysler called its new van a "family room on wheels."

▼ *Owners of sport-utility vehicles (SUVs) are affected by high gas prices and shortages because the fuel efficiency of SUVs is quite low compared to that of other vehicles.*

During the 1990s, another vehicle became popular with families: the sport-utility vehicle (SUV). The first SUVs were made by a company called American Motors. These vehicles traced their roots to the Jeep, a small vehicle developed for the U.S. military during World War II. The first Jeeps were designed to travel off paved roads, through mud, and over rocks and fallen trees. The American Motors Jeep Cherokee was an SUV. It combined a passenger car with the lower body of a truck. Drivers could take their Cherokees off roads and onto dirt trails. The vehicles also had **four-wheel drive**. This feature made it easier for them to drive through snow and mud.

By 2001, one out of every four cars sold in the United States was an SUV. Most people used them as family cars. Like minivans, larger SUVs had room for cargo, as well as passengers.

Some people, however, disliked SUVs. These critics pointed out that SUVs had much worse gas mileage than regular cars and produced more pollution. By selling SUVs, car companies avoided U.S. laws regarding gas mileage for cars—technically, the SUVs were light trucks. Trucks were allowed to have lower fuel mileage

Fast Fact

Chryslers remain the world's best-selling minivans, and the company that produces them also makes Jeeps.

Danger on Four Wheels?

During the 1980s, the Ford Motor Company introduced its first popular SUV, the Bronco II. Tests showed that the vehicle could tip over at speeds as low as 20 miles (32 km) per hour. Ford knew this, yet did not take steps to make the car safer. The U.S. government also refused to limit what kind of SUVs were sold, even though it knew the vehicles could be unsafe. One former government official who opposed limits on SUVs said in 2001, "The American people are not stupid. They buy what they want to buy. They know that vehicle is higher [off the ground]. And if in all of the TV press you've gotten over the last 20 years, you don't now know that an SUV is not like a car, something is wrong with you!" He meant that people had to realize cars and SUVS were not the same. People had to be more careful when they drove SUVs. Defenders of SUVs also said that people were more likely to survive a crash in an SUV than in a regular car.

than passenger cars. SUV critics also noted that the vehicles could be dangerous. Tests showed that SUVs were more likely to tip over than cars. They also sometimes blocked the views of people driving smaller cars. SUVs could heavily damage smaller cars if they hit them during accidents, as well. Many people, however, continued to buy SUVs, and some companies introduced much larger models than the first Jeeps.

The Gas Issue—Again

SUVs and other large vehicles have been called "gas guzzlers" because they use more gas than the average passenger car. Improving gas mileage is a continuing concern for auto engineers. Some of the most important automotive research has gone into building cars that use less gas or none at all.

One answer seemed to be electric cars. Early electric cars had failed because they could not travel far without recharging their batteries. In 1996, GM introduced the first modern electric car for everyday use. Its EV1 could go several hundred miles before it needed a recharge. The car was not meant for long trips, but for commuters who liked the idea of a fast, silent car that did not use gas or pollute. Not enough people were interested in driving electric cars for GM to keep making them, however. The company took the remaining cars off the road in 2003.

By that time, electric and gas motors had been combined in cars called hybrids. A hybrid draws power from both of its motors at the same time. At low speeds or while stopped, just the electric motor works. When the car needs extra power, the gasoline engine begins to operate, too. The battery for the electric motor recharges while the car is running. Energy created when the brakes are pressed is sent to the electric motor. In that instant, the motor begins to act like a generator and sends electric power back to the battery. The hybrid never has to be plugged in to recharge as the EV1 and earlier electric cars did.

The first hybrids were small passengers cars. They were able to go more than 50 miles (80 km) on 1 gallon (3.8 l) of gas. They also produced less air pollution than the average passenger car. In 2004, the first hybrid pickup trucks and SUVs appeared. Thanks to their electric motors, they had much higher gas mileage than standard pickups and SUVs. Their gasoline engines gave them plenty of power and speed for highway driving or going up hills.

Fast Fact

By 2004, light trucks—including SUVs—made up sixty percent of all vehicles sold in the United States.

▲ *GM's EV1 electric car was attractive and efficient, but failed to catch on with a public that, ten years ago, was even less concerned with alternate forms of fuel than it is today.*

Fast Fact

General Motors spent more than $1 billion designing and making the EV1 electric car.

▼ *This fuel cell–powered car made by Humbolt University Schatz Energy Research Center is a prototype for future advances in fuel cell technology.*

Looking to the Future

Hybrids will become more common in the years to come. Some auto engineers hope another kind of power will also reduce the need for gasoline-powered engines. A source of energy called a fuel cell has existed for more than 150 years. Fuel cells are similar to batteries. They have a positive and a negative end. Fuel cells mix oxygen and hydrogen. Combining these two gases creates electricity and water. One fuel cell cannot create much electricity, so many of them are combined into what are called stacks. A fuel cell stack can create enough electricity to power an electric motor in a vehicle.

Fuel cells have several advantages over regular batteries. They do not have to be recharged. As long as they are supplied with hydrogen mixed with oxygen, they can create power. Hydrogen is plentiful, so people will never run out of it. Fuel cells are also clean—they do not pollute. Motors that run on fuel cells are already being used to power buses and some cars.

Building a fuel cell–powered car, however, presents problems. Fuel cells small enough and powerful enough for cars are expensive. Engineers have to find ways to build them more cheaply so most people can afford them. Hydrogen is also expensive to use as a fuel. Hydrogen fuel can explode easily. Carmakers have to guarantee that the fuel would not pose a danger if a fuel cell car crashed. Despite these challenges, fuelcells offer great promise. Still, some engineers say cars will continue to use gasoline for decades to come.

No matter how their cars are powered, carmakers will focus on making them safer. Traffic accidents still kill more than 40,000 Americans each year. Injuries from car accidents, however, have fallen slightly, because more cars have better protection for passengers during crashes. Front air bags are common on almost all new cars, and many models also have bags that protect passengers from side crashes.

Automakers are also testing systems that use computers and video cameras to help drivers avoid accidents. One system uses a laser to help a video camera see the road ahead. The laser helps the camera detect images before drivers can see them with their eyes.

Computer systems are also being used to share information between cars and the roads on which they travel. Sensors along these "intelligent highways" gather information on the weather and traffic conditions. Some cars also have similar sensors. Electronic signs along the road can alert drivers to bad conditions ahead. In the future, drivers may get these warnings on cell phones or other wireless communication systems.

Some intelligent highways work with cars that have global positioning systems (GPS) built into them. Satellites circling Earth can pinpoint the location of cars that have GPS. Global positioning helps rescuers locate cars that have been in accidents in remote locations. With a GPS location, rescuers can reach the accident scene faster and possibly save the life of someone injured in a crash.

Finding new ways to use computers should help the world's automakers build cars that are safer and easier to drive. The auto has changed how people travel and live. Although cars have created some challenges and problems, it's hard to imagine the modern world without them.

Power from the Sun

In recent years, some engineers have built experimental cars powered by the Sun. The car shown above, built by the Rose-Hulman Institute of Technology, has small devices called solar cells that turn energy from the Sun into electricity. This electricity recharges batteries that power an electric motor. Like hydrogen-powered cars, solar cars do not produce pollution. They also use an energy supply that will never run out. Solar-powered cars have some problems, however. Current models cannot travel very fast, and most carry just a driver. The cars also lose energy on cloudy days. Over time, however, improved solar cells may lead to practical solar cars.

TIME LINE

c. 3500 B.C.	The wheel is invented.
c. A.D. 500	Chinese engineers build wind-powered cars.
1712	Thomas Newcomen develops a working steam engine.
1769	Nicolas-Joseph Cugnot creates the first steam-powered road vehicle.
1850s	Scientists discover that petroleum can be refined into different products, including gasoline.
1876	Nikolaus Otto perfects the four-stroke internal **combustion** engine.
1885	Gottlieb Daimler builds a successful gasoline-powered engine and makes the first motorcycle; Carl Benz introduces the first practical gasoline-powered car.
1888	John Dunlop patents the pneumatic tire.
1890	The first practical electric cars appear.
1896	The first recorded U.S. traffic accident occurs.
1901	Connecticut is the first U.S. state to post speed limits.
1910s	Improvements to cars include electric headlights, starters, and all-wheel brakes.
1913	Henry Ford perfects mass production.
1930s	Seat belts appear for the first time.
1946–1950	Americans buy more than 21 million vehicles.
1952	John Hetrick invents the first air bag.
1966	The U.S. Congress passes a law calling for safer cars.
1970	The U.S. Congress calls for cars that produce less air pollution.
1970s	Carmakers add computers to cars to improve gas mileage and overall performance.
1973	The Arab oil embargo limits the supply of available gasoline, which raises the price of gas.
1983	Chrysler introduces the first minivan.
1996	General Motors introduces the EV1, the first modern electric car.
1997	Japanese automakers Toyota and Honda begin selling hybrid cars, which use both gasoline engines and electric motors.
2001	One out of every four cars sold in the United States is a sport-utility vehicle (SUV).
2003	U.S. president George W. Bush requests money to build cars that will run on hydrogen.
2004	The first hybrid SUVs and pickup trucks go on sale.

GLOSSARY

advocate: a person who looks out for the public interests of others

antique: from a former period

civilians: people not in the military

commuted: traveled between work and home

defective: broken or not working properly

electromagnet: a magnet created by sending electricity through a metal coil

four-wheel drive: a vehicle driving system that sends power to all four wheels, increasing traction

horsepower (hp): the unit of measure for the energy created by an engine; 1 hp equals 33,000 foot-pounds (44,750 joules) per minute

hydraulic: relating to the pressure created by fluids in a sealed system

ignited: lit by a spark to start a fire or explosion

internal combustion: the burning of a fuel inside something, such as the cylinder of a car engine

interstate: of, existing in, or connecting two or more states

patent: a legal document that prevents people from stealing or making money from another person's idea or invention

pedestrians: people who walk on or along a street

piston: a solid cylinder that fits snugly into a larger cylinder and moves under fluid pressure

pneumatic: relating to air pressure

rationed: sold in limited amounts

refining: the process of separating petroleum into different products

robotic: describing something similar to a robot, a mechanical device that performs work usually done by people

trolleys: used to carry passengers along city streets, they were usually powered by over head electric wires; like trains, they ran on rails

self-propelled: able to move by itself

unions: groups of workers formed to improve pay and working conditions in an industry

vaporize: to turn into a powder or mist

FOR MORE INFORMATION

Books

Burgan, Michael. *Henry Ford.* Chicago: Ferguson Publishing, 2001.

Conley, Robyn, and F. Watts. *The Automobile.* Inventions That Shaped the World. New York: Scholastic Library Publishing, 2005.

Ditchfield, Christin. *Oil.* New York: Children's Press, 2002.

Flammang, James M. *Cars.* Berkeley Heights, NJ: Enslow Publishers, 2001.

Italia, Bob. *Great Auto Makers and Their Cars.* Minneapolis: Oliver Press, 1993.

Sutton, Richard. *Car.* New York: Dorling Kindersley, 2000.

Videos and DVDs

America on Wheels. Atlanta: Turner Learning, 1997.

Dream Cars. New York: A&E Home Video, 1998.

Electric Cars. Santa Monica, CA: Main Street Media, 2000.

How Are Cars Made? Roeland Park, KS: Oak Leaf Productions, 2000.

Web Sites

americanhistory.si.edu/onthemove/ This Web site explores the way transportation shaped the history and culture of the United States.

auto.howstuffworks.com Viewers learn how all parts of a car work, in addition to the practical issues attached to buying and owning a car.

msnbc.com/modules/summer_driving/decades/default.asp Enjoy humorous news story about vacation road trips through the years and how to survive them.

www.automotivehalloffame.com An interactive Web site that allows students to design imaginary cars, see videos of future transportation, and explore the evolution of the automobile.

www.thehenryford.org/explore/exhibits.asp This Web site for the Henry Ford Museum features exhibits that honor Americans whose ideas have changed our lives.

INDEX

Author Biography

As an editor at *Weekly Reader* for six years, Michael Burgan created educational material for an interactive online service and wrote on current events. Now a freelance author, Michael has written more than ninety books for children and young adults. These include the five-volume set *Science in Everyday Life in America* and two biographies of Henry Ford. Michael has a B.A. in history from the University of Connecticut. He resides in Chicago.